THIS BOOK BELONGS TO:

CONTACT INFORMATION	
NAME	
PHONE #	
EMAIL	

DEDICATION

This Bug identification Log Book is dedicated to kids who love bugs, nature, camping, and other outdoor activities.

You are my inspiration for producing this book and I'm honored to be a part of your record-keeping and organization.

HOW TO USE THIS BOOK

This Bug Identification Log Book will help you by accurately recording, and organizing your information in an easy to use format.

Here are examples of information for you to fill in and write the details of your logbook.

Fill in the following information:

1. Date, Time, and Season — Record information.

2. Weather Conditions — A checklist to record the day's weather conditions.

3. Bug Name — Write down the name of the bug.

4. Scientific Name — Do some research at home and write down the scientific name of the bug.

5. Where Did You Find The Bug? — Fill in the location.

6. Number of Legs — Examine the bug and record the number of legs.

7. Does The Bug Have Wings? — Checklist (yes, no, not sure)

8. The Bug Is — Use the checklist to note the characteristics of the bug (big, shiny, scary, little, etc.).

9. Does The Bug Make a Sound? — Checklist (yes or no)

10. Was It In A Group or Was It Alone? — Checklist (yes or no)

11. Notes — Use this space to write notes regarding the bug (habitat, what you were doing when you found the bug, research, and other information).

12. Photo/Drawing — Paste a photo or create a drawing of the bug found.

BUG IDENTIFICATION

DATE:		TIME:		SEASON:	○ SPRING ○ SUMMER ○ FALL ○ WINTER

WEATHER CONDITIONS:	○ HOT ○ WARM ○ SUNNY ○ CLOUDY ○ RAINY ○ WINDY ○ FOGGY ○ COLD
BUG NAME:	
SCIENTIFIC NAME:	
WHERE DID YOU FIND IT?	
WHAT COLOR(S) IS THE BUG?	

NUMBER OF LEGS?		DOES IT HAVE WINGS?	○ YES ○ NO ○ NOT SURE

THE BUG IS...	○ BIG ○ SHINY ○ FAST ○ SCARY ○ LITTLE ○ SLOW ○ CUTE ○ ROUND ○ THIN

DOES IT MAKE ANY SOUND?	○ YES ○ NO	WAS IT ALONE OR IN A GROUP?	○ ALONE ○ GROUP

NOTES

PHOTO/ DRAWING

BUG IDENTIFICATION

DATE:		TIME:		SEASON:	○ SPRING ○ SUMMER ○ FALL ○ WINTER		
WEATHER CONDITIONS:		○ HOT ○ WARM ○ SUNNY ○ CLOUDY ○ RAINY ○ WINDY ○ FOGGY ○ COLD					
BUG NAME:							
SCIENTIFIC NAME:							
WHERE DID YOU FIND IT?							
WHAT COLOR(S) IS THE BUG?							
NUMBER OF LEGS?			DOES IT HAVE WINGS?		○ YES ○ NO ○ NOT SURE		
THE BUG IS...		○ BIG ○ SHINY ○ FAST ○ SCARY ○ LITTLE ○ SLOW ○ CUTE ○ ROUND ○ THIN					
DOES IT MAKE ANY SOUND?		○ YES ○ NO	WAS IT ALONE OR IN A GROUP?			○ ALONE ○ GROUP	

NOTES

PHOTO/DRAWING

BUG IDENTIFICATION

DATE:		TIME:		SEASON:	○ SPRING ○ SUMMER ○ FALL ○ WINTER

WEATHER CONDITIONS:	○ HOT ○ WARM ○ SUNNY ○ CLOUDY ○ RAINY ○ WINDY ○ FOGGY ○ COLD

BUG NAME:	
SCIENTIFIC NAME:	
WHERE DID YOU FIND IT?	
WHAT COLOR(S) IS THE BUG?	

NUMBER OF LEGS?		DOES IT HAVE WINGS?	○ YES ○ NO ○ NOT SURE

THE BUG IS...	○ BIG ○ SHINY ○ FAST ○ SCARY ○ LITTLE ○ SLOW ○ CUTE ○ ROUND ○ THIN

DOES IT MAKE ANY SOUND?	○ YES ○ NO	WAS IT ALONE OR IN A GROUP?	○ ALONE ○ GROUP

NOTES

PHOTO / DRAWING

BUG IDENTIFICATION

DATE:		TIME:		SEASON:	○ SPRING ○ SUMMER ○ FALL ○ WINTER
WEATHER CONDITIONS:		○ HOT ○ WARM ○ SUNNY ○ CLOUDY ○ RAINY ○ WINDY ○ FOGGY ○ COLD				
BUG NAME:					
SCIENTIFIC NAME:					
WHERE DID YOU FIND IT?					
WHAT COLOR(S) IS THE BUG?					
NUMBER OF LEGS?		**DOES IT HAVE WINGS?**		○ YES ○ NO ○ NOT SURE	
THE BUG IS...		○ BIG ○ SHINY ○ FAST ○ SCARY ○ LITTLE ○ SLOW ○ CUTE ○ ROUND ○ THIN				
DOES IT MAKE ANY SOUND?	○ YES ○ NO	**WAS IT ALONE OR IN A GROUP?**		○ ALONE ○ GROUP	

NOTES

PHOTO / DRAWING

BUG IDENTIFICATION

DATE:		TIME:		SEASON:	○ SPRING ○ SUMMER ○ FALL ○ WINTER			
WEATHER CONDITIONS:		○ HOT ○ WARM ○ SUNNY ○ CLOUDY ○ RAINY ○ WINDY ○ FOGGY ○ COLD						
BUG NAME:								
SCIENTIFIC NAME:								
WHERE DID YOU FIND IT?								
WHAT COLOR(S) IS THE BUG?								
NUMBER OF LEGS?		**DOES IT HAVE WINGS?**		○ YES ○ NO ○ NOT SURE				
THE BUG IS...		○ BIG ○ SHINY ○ FAST ○ SCARY ○ LITTLE ○ SLOW ○ CUTE ○ ROUND ○ THIN						
DOES IT MAKE ANY SOUND?	○ YES ○ NO	**WAS IT ALONE OR IN A GROUP?**		○ ALONE ○ GROUP				

NOTES

PHOTO / DRAWING

BUG IDENTIFICATION

DATE:		TIME:		SEASON:	○ SPRING ○ SUMMER ○ FALL ○ WINTER		
WEATHER CONDITIONS:		○ HOT ○ WARM ○ SUNNY ○ CLOUDY ○ RAINY ○ WINDY ○ FOGGY ○ COLD					
BUG NAME:							
SCIENTIFIC NAME:							
WHERE DID YOU FIND IT?							
WHAT COLOR(S) IS THE BUG?							
NUMBER OF LEGS?			DOES IT HAVE WINGS?		○ YES ○ NO ○ NOT SURE		
THE BUG IS...		○ BIG ○ SHINY ○ FAST ○ SCARY ○ LITTLE ○ SLOW ○ CUTE ○ ROUND ○ THIN					
DOES IT MAKE ANY SOUND?	○ YES ○ NO		WAS IT ALONE OR IN A GROUP?		○ ALONE ○ GROUP		

NOTES

PHOTO / DRAWING

BUG IDENTIFICATION

DATE:		TIME:		SEASON:	○ SPRING ○ SUMMER ○ FALL ○ WINTER		
WEATHER CONDITIONS:		○ HOT ○ WARM ○ SUNNY ○ CLOUDY ○ RAINY ○ WINDY ○ FOGGY ○ COLD					
BUG NAME:							
SCIENTIFIC NAME:							
WHERE DID YOU FIND IT?							
WHAT COLOR(S) IS THE BUG?							
NUMBER OF LEGS?			DOES IT HAVE WINGS?		○ YES ○ NO ○ NOT SURE		
THE BUG IS...		○ BIG ○ SHINY ○ FAST ○ SCARY ○ LITTLE ○ SLOW ○ CUTE ○ ROUND ○ THIN					
DOES IT MAKE ANY SOUND?		○ YES ○ NO	WAS IT ALONE OR IN A GROUP?			○ ALONE ○ GROUP	

NOTES

PHOTO / DRAWING

BUG IDENTIFICATION

DATE:		TIME:		SEASON:	○ SPRING ○ SUMMER ○ FALL ○ WINTER		
WEATHER CONDITIONS:		○ HOT ○ WARM ○ SUNNY ○ CLOUDY ○ RAINY ○ WINDY ○ FOGGY ○ COLD					
BUG NAME:							
SCIENTIFIC NAME:							
WHERE DID YOU FIND IT?							
WHAT COLOR(S) IS THE BUG?							
NUMBER OF LEGS?		DOES IT HAVE WINGS?		○ YES ○ NO ○ NOT SURE			
THE BUG IS...		○ BIG ○ SHINY ○ FAST ○ SCARY ○ LITTLE ○ SLOW ○ CUTE ○ ROUND ○ THIN					
DOES IT MAKE ANY SOUND?	○ YES ○ NO	WAS IT ALONE OR IN A GROUP?		○ ALONE ○ GROUP			

NOTES

PHOTO / DRAWING

BUG IDENTIFICATION

DATE:		TIME:		SEASON:	○ SPRING ○ SUMMER ○ FALL ○ WINTER

WEATHER CONDITIONS:	○ HOT ○ WARM ○ SUNNY ○ CLOUDY ○ RAINY ○ WINDY ○ FOGGY ○ COLD
BUG NAME:	
SCIENTIFIC NAME:	
WHERE DID YOU FIND IT?	
WHAT COLOR(S) IS THE BUG?	

NUMBER OF LEGS?		DOES IT HAVE WINGS?	○ YES ○ NO ○ NOT SURE

THE BUG IS...	○ BIG ○ SHINY ○ FAST ○ SCARY ○ LITTLE ○ SLOW ○ CUTE ○ ROUND ○ THIN

DOES IT MAKE ANY SOUND?	○ YES ○ NO	WAS IT ALONE OR IN A GROUP?	○ ALONE ○ GROUP

NOTES

PHOTO / DRAWING

BUG IDENTIFICATION

DATE:		TIME:		SEASON:	○ SPRING ○ SUMMER ○ FALL ○ WINTER
WEATHER CONDITIONS:		○ HOT ○ WARM ○ SUNNY ○ CLOUDY ○ RAINY ○ WINDY ○ FOGGY ○ COLD			
BUG NAME:					
SCIENTIFIC NAME:					
WHERE DID YOU FIND IT?					
WHAT COLOR(S) IS THE BUG?					
NUMBER OF LEGS?		**DOES IT HAVE WINGS?**		○ YES ○ NO ○ NOT SURE	
THE BUG IS...		○ BIG ○ SHINY ○ FAST ○ SCARY ○ LITTLE ○ SLOW ○ CUTE ○ ROUND ○ THIN			
DOES IT MAKE ANY SOUND?	○ YES ○ NO	**WAS IT ALONE OR IN A GROUP?**		○ ALONE ○ GROUP	

NOTES

PHOTO/DRAWING

BUG IDENTIFICATION

DATE:		TIME:		SEASON:	○ SPRING ○ SUMMER ○ FALL ○ WINTER
WEATHER CONDITIONS:		○ HOT ○ WARM ○ SUNNY ○ CLOUDY ○ RAINY ○ WINDY ○ FOGGY ○ COLD			
BUG NAME:					
SCIENTIFIC NAME:					
WHERE DID YOU FIND IT?					
WHAT COLOR(S) IS THE BUG?					
NUMBER OF LEGS?		**DOES IT HAVE WINGS?**		○ YES ○ NO ○ NOT SURE	
THE BUG IS...		○ BIG ○ SHINY ○ FAST ○ SCARY ○ LITTLE ○ SLOW ○ CUTE ○ ROUND ○ THIN			
DOES IT MAKE ANY SOUND?	○ YES ○ NO	**WAS IT ALONE OR IN A GROUP?**		○ ALONE ○ GROUP	

NOTES

PHOTO / DRAWING

BUG IDENTIFICATION

DATE:		TIME:		SEASON:	○ SPRING ○ SUMMER ○ FALL ○ WINTER

WEATHER CONDITIONS:	○ HOT ○ WARM ○ SUNNY ○ CLOUDY ○ RAINY ○ WINDY ○ FOGGY ○ COLD

BUG NAME:	
SCIENTIFIC NAME:	
WHERE DID YOU FIND IT?	
WHAT COLOR(S) IS THE BUG?	

NUMBER OF LEGS?		DOES IT HAVE WINGS?	○ YES ○ NO ○ NOT SURE

THE BUG IS...	○ BIG ○ SHINY ○ FAST ○ SCARY ○ LITTLE ○ SLOW ○ CUTE ○ ROUND ○ THIN

DOES IT MAKE ANY SOUND?	○ YES ○ NO	WAS IT ALONE OR IN A GROUP?	○ ALONE ○ GROUP

NOTES

PHOTO / DRAWING

BUG IDENTIFICATION

DATE:		TIME:		SEASON:	○ SPRING ○ SUMMER ○ FALL ○ WINTER

WEATHER CONDITIONS:	○ HOT ○ WARM ○ SUNNY ○ CLOUDY ○ RAINY ○ WINDY ○ FOGGY ○ COLD
BUG NAME:	
SCIENTIFIC NAME:	
WHERE DID YOU FIND IT?	
WHAT COLOR(S) IS THE BUG?	

NUMBER OF LEGS?		DOES IT HAVE WINGS?		○ YES ○ NO ○ NOT SURE

THE BUG IS...	○ BIG ○ SHINY ○ FAST ○ SCARY ○ LITTLE ○ SLOW ○ CUTE ○ ROUND ○ THIN

DOES IT MAKE ANY SOUND?	○ YES ○ NO	WAS IT ALONE OR IN A GROUP?	○ ALONE ○ GROUP

NOTES

PHOTO / DRAWING

BUG IDENTIFICATION

DATE:		TIME:		SEASON:	○ SPRING ○ SUMMER ○ FALL ○ WINTER			
WEATHER CONDITIONS:		○ HOT ○ WARM ○ SUNNY ○ CLOUDY ○ RAINY ○ WINDY ○ FOGGY ○ COLD						
BUG NAME:								
SCIENTIFIC NAME:								
WHERE DID YOU FIND IT?								
WHAT COLOR(S) IS THE BUG?								
NUMBER OF LEGS?		**DOES IT HAVE WINGS?**			○ YES ○ NO ○ NOT SURE			
THE BUG IS...		○ BIG ○ SHINY ○ FAST ○ SCARY ○ LITTLE ○ SLOW ○ CUTE ○ ROUND ○ THIN						
DOES IT MAKE ANY SOUND?		○ YES ○ NO	**WAS IT ALONE OR IN A GROUP?**			○ ALONE ○ GROUP		

NOTES

PHOTO/DRAWING

BUG IDENTIFICATION

DATE:		TIME:		SEASON:	○ SPRING ○ SUMMER ○ FALL ○ WINTER
WEATHER CONDITIONS:		○ HOT ○ WARM ○ SUNNY ○ CLOUDY ○ RAINY ○ WINDY ○ FOGGY ○ COLD			
BUG NAME:					
SCIENTIFIC NAME:					
WHERE DID YOU FIND IT?					
WHAT COLOR(S) IS THE BUG?					
NUMBER OF LEGS?		**DOES IT HAVE WINGS?**		○ YES ○ NO ○ NOT SURE	
THE BUG IS...		○ BIG ○ SHINY ○ FAST ○ SCARY ○ LITTLE ○ SLOW ○ CUTE ○ ROUND ○ THIN			
DOES IT MAKE ANY SOUND?	○ YES ○ NO	**WAS IT ALONE OR IN A GROUP?**		○ ALONE ○ GROUP	

NOTES

PHOTO / DRAWING

BUG IDENTIFICATION

DATE:		TIME:		SEASON:	○ SPRING ○ SUMMER ○ FALL ○ WINTER		
WEATHER CONDITIONS:		○ HOT ○ WARM ○ SUNNY ○ CLOUDY ○ RAINY ○ WINDY ○ FOGGY ○ COLD					
BUG NAME:							
SCIENTIFIC NAME:							
WHERE DID YOU FIND IT?							
WHAT COLOR(S) IS THE BUG?							
NUMBER OF LEGS?			DOES IT HAVE WINGS?		○ YES ○ NO ○ NOT SURE		
THE BUG IS...		○ BIG ○ SHINY ○ FAST ○ SCARY ○ LITTLE ○ SLOW ○ CUTE ○ ROUND ○ THIN					
DOES IT MAKE ANY SOUND?		○ YES ○ NO	WAS IT ALONE OR IN A GROUP?			○ ALONE ○ GROUP	

NOTES

PHOTO / DRAWING

BUG IDENTIFICATION

DATE:		TIME:		SEASON:	○ SPRING ○ SUMMER ○ FALL ○ WINTER

WEATHER CONDITIONS:	○ HOT ○ WARM ○ SUNNY ○ CLOUDY ○ RAINY ○ WINDY ○ FOGGY ○ COLD

BUG NAME:	
SCIENTIFIC NAME:	
WHERE DID YOU FIND IT?	
WHAT COLOR(S) IS THE BUG?	

NUMBER OF LEGS?		DOES IT HAVE WINGS?	○ YES ○ NO ○ NOT SURE

THE BUG IS...	○ BIG ○ SHINY ○ FAST ○ SCARY ○ LITTLE ○ SLOW ○ CUTE ○ ROUND ○ THIN

DOES IT MAKE ANY SOUND?	○ YES ○ NO	WAS IT ALONE OR IN A GROUP?	○ ALONE ○ GROUP

NOTES

PHOTO / DRAWING

BUG IDENTIFICATION

DATE:		TIME:		SEASON:	○ SPRING ○ SUMMER ○ FALL ○ WINTER		
WEATHER CONDITIONS:		○ HOT ○ WARM ○ SUNNY ○ CLOUDY ○ RAINY ○ WINDY ○ FOGGY ○ COLD					
BUG NAME:							
SCIENTIFIC NAME:							
WHERE DID YOU FIND IT?							
WHAT COLOR(S) IS THE BUG?							
NUMBER OF LEGS?			DOES IT HAVE WINGS?		○ YES ○ NO ○ NOT SURE		
THE BUG IS...		○ BIG ○ SHINY ○ FAST ○ SCARY ○ LITTLE ○ SLOW ○ CUTE ○ ROUND ○ THIN					
DOES IT MAKE ANY SOUND?		○ YES ○ NO	WAS IT ALONE OR IN A GROUP?			○ ALONE ○ GROUP	

NOTES

PHOTO / DRAWING

BUG IDENTIFICATION

DATE:		TIME:		SEASON:	○ SPRING ○ SUMMER ○ FALL ○ WINTER

WEATHER CONDITIONS:	○ HOT ○ WARM ○ SUNNY ○ CLOUDY ○ RAINY ○ WINDY ○ FOGGY ○ COLD

BUG NAME:	
SCIENTIFIC NAME:	
WHERE DID YOU FIND IT?	
WHAT COLOR(S) IS THE BUG?	

NUMBER OF LEGS?		DOES IT HAVE WINGS?		○ YES ○ NO ○ NOT SURE

THE BUG IS...	○ BIG ○ SHINY ○ FAST ○ SCARY ○ LITTLE ○ SLOW ○ CUTE ○ ROUND ○ THIN

DOES IT MAKE ANY SOUND?	○ YES ○ NO	WAS IT ALONE OR IN A GROUP?	○ ALONE ○ GROUP

NOTES

PHOTO / DRAWING

BUG IDENTIFICATION

DATE:		TIME:		SEASON:	○ SPRING ○ SUMMER ○ FALL ○ WINTER

WEATHER CONDITIONS:	○ HOT ○ WARM ○ SUNNY ○ CLOUDY ○ RAINY ○ WINDY ○ FOGGY ○ COLD
BUG NAME:	
SCIENTIFIC NAME:	
WHERE DID YOU FIND IT?	
WHAT COLOR(S) IS THE BUG?	

NUMBER OF LEGS?		DOES IT HAVE WINGS?	○ YES ○ NO ○ NOT SURE

THE BUG IS...	○ BIG ○ SHINY ○ FAST ○ SCARY ○ LITTLE ○ SLOW ○ CUTE ○ ROUND ○ THIN

DOES IT MAKE ANY SOUND?	○ YES ○ NO	WAS IT ALONE OR IN A GROUP?	○ ALONE ○ GROUP

NOTES

PHOTO / DRAWING

BUG IDENTIFICATION

DATE:		TIME:		SEASON:	○ SPRING ○ SUMMER ○ FALL ○ WINTER

WEATHER CONDITIONS:	○ HOT ○ WARM ○ SUNNY ○ CLOUDY ○ RAINY ○ WINDY ○ FOGGY ○ COLD
BUG NAME:	
SCIENTIFIC NAME:	
WHERE DID YOU FIND IT?	
WHAT COLOR(S) IS THE BUG?	

NUMBER OF LEGS?		DOES IT HAVE WINGS?	○ YES ○ NO ○ NOT SURE

THE BUG IS...	○ BIG ○ SHINY ○ FAST ○ SCARY ○ LITTLE ○ SLOW ○ CUTE ○ ROUND ○ THIN

DOES IT MAKE ANY SOUND?	○ YES ○ NO	WAS IT ALONE OR IN A GROUP?	○ ALONE ○ GROUP

NOTES

PHOTO / DRAWING

BUG IDENTIFICATION

DATE:		TIME:		SEASON:	○ SPRING ○ SUMMER ○ FALL ○ WINTER

WEATHER CONDITIONS:	○ HOT ○ WARM ○ SUNNY ○ CLOUDY ○ RAINY ○ WINDY ○ FOGGY ○ COLD

BUG NAME:	
SCIENTIFIC NAME:	
WHERE DID YOU FIND IT?	
WHAT COLOR(S) IS THE BUG?	

NUMBER OF LEGS?		DOES IT HAVE WINGS?	○ YES ○ NO ○ NOT SURE

THE BUG IS...	○ BIG ○ SHINY ○ FAST ○ SCARY ○ LITTLE ○ SLOW ○ CUTE ○ ROUND ○ THIN

DOES IT MAKE ANY SOUND?	○ YES ○ NO	WAS IT ALONE OR IN A GROUP?	○ ALONE ○ GROUP

NOTES

PHOTO / DRAWING

BUG IDENTIFICATION

DATE:		TIME:		SEASON:	○ SPRING ○ SUMMER ○ FALL ○ WINTER

WEATHER CONDITIONS:	○ HOT ○ WARM ○ SUNNY ○ CLOUDY ○ RAINY ○ WINDY ○ FOGGY ○ COLD

BUG NAME:	
SCIENTIFIC NAME:	
WHERE DID YOU FIND IT?	
WHAT COLOR(S) IS THE BUG?	

NUMBER OF LEGS?		DOES IT HAVE WINGS?	○ YES ○ NO ○ NOT SURE

THE BUG IS...	○ BIG ○ SHINY ○ FAST ○ SCARY ○ LITTLE ○ SLOW ○ CUTE ○ ROUND ○ THIN

DOES IT MAKE ANY SOUND?	○ YES ○ NO	WAS IT ALONE OR IN A GROUP?	○ ALONE ○ GROUP

NOTES

PHOTO / DRAWING

BUG IDENTIFICATION

DATE:		TIME:		SEASON:	○ SPRING ○ SUMMER ○ FALL ○ WINTER

WEATHER CONDITIONS:	○ HOT ○ WARM ○ SUNNY ○ CLOUDY ○ RAINY ○ WINDY ○ FOGGY ○ COLD
BUG NAME:	
SCIENTIFIC NAME:	
WHERE DID YOU FIND IT?	
WHAT COLOR(S) IS THE BUG?	

NUMBER OF LEGS?		DOES IT HAVE WINGS?	○ YES ○ NO ○ NOT SURE

THE BUG IS...	○ BIG ○ SHINY ○ FAST ○ SCARY ○ LITTLE ○ SLOW ○ CUTE ○ ROUND ○ THIN

DOES IT MAKE ANY SOUND?	○ YES ○ NO	WAS IT ALONE OR IN A GROUP?	○ ALONE ○ GROUP

NOTES

PHOTO / DRAWING

BUG IDENTIFICATION

DATE:		TIME:		SEASON:	○ SPRING ○ SUMMER ○ FALL ○ WINTER

WEATHER CONDITIONS:	○ HOT ○ WARM ○ SUNNY ○ CLOUDY ○ RAINY ○ WINDY ○ FOGGY ○ COLD

BUG NAME:	
SCIENTIFIC NAME:	
WHERE DID YOU FIND IT?	
WHAT COLOR(S) IS THE BUG?	

NUMBER OF LEGS?		DOES IT HAVE WINGS?	○ YES ○ NO ○ NOT SURE

THE BUG IS...	○ BIG ○ SHINY ○ FAST ○ SCARY ○ LITTLE ○ SLOW ○ CUTE ○ ROUND ○ THIN

DOES IT MAKE ANY SOUND?	○ YES ○ NO	WAS IT ALONE OR IN A GROUP?	○ ALONE ○ GROUP

NOTES

PHOTO / DRAWING

BUG IDENTIFICATION

DATE:		TIME:		SEASON:	○ SPRING ○ SUMMER ○ FALL ○ WINTER

WEATHER CONDITIONS:	○ HOT ○ WARM ○ SUNNY ○ CLOUDY ○ RAINY ○ WINDY ○ FOGGY ○ COLD
BUG NAME:	
SCIENTIFIC NAME:	
WHERE DID YOU FIND IT?	
WHAT COLOR(S) IS THE BUG?	

NUMBER OF LEGS?		DOES IT HAVE WINGS?	○ YES ○ NO ○ NOT SURE

THE BUG IS...	○ BIG ○ SHINY ○ FAST ○ SCARY ○ LITTLE ○ SLOW ○ CUTE ○ ROUND ○ THIN

DOES IT MAKE ANY SOUND?	○ YES ○ NO	WAS IT ALONE OR IN A GROUP?	○ ALONE ○ GROUP

NOTES

PHOTO/DRAWING

BUG IDENTIFICATION

DATE:		TIME:		SEASON:	○ SPRING ○ SUMMER ○ FALL ○ WINTER		
WEATHER CONDITIONS:		○ HOT ○ WARM ○ SUNNY ○ CLOUDY ○ RAINY ○ WINDY ○ FOGGY ○ COLD					
BUG NAME:							
SCIENTIFIC NAME:							
WHERE DID YOU FIND IT?							
WHAT COLOR(S) IS THE BUG?							
NUMBER OF LEGS?		**DOES IT HAVE WINGS?**		○ YES ○ NO ○ NOT SURE			
THE BUG IS...		○ BIG ○ SHINY ○ FAST ○ SCARY ○ LITTLE ○ SLOW ○ CUTE ○ ROUND ○ THIN					
DOES IT MAKE ANY SOUND?	○ YES ○ NO	**WAS IT ALONE OR IN A GROUP?**		○ ALONE ○ GROUP			

NOTES

PHOTO / DRAWING

BUG IDENTIFICATION

DATE:		TIME:		SEASON:	○ SPRING ○ SUMMER ○ FALL ○ WINTER		
WEATHER CONDITIONS:		○ HOT ○ WARM ○ SUNNY ○ CLOUDY ○ RAINY ○ WINDY ○ FOGGY ○ COLD					
BUG NAME:							
SCIENTIFIC NAME:							
WHERE DID YOU FIND IT?							
WHAT COLOR(S) IS THE BUG?							
NUMBER OF LEGS?		DOES IT HAVE WINGS?		○ YES ○ NO ○ NOT SURE			
THE BUG IS...		○ BIG ○ SHINY ○ FAST ○ SCARY ○ LITTLE ○ SLOW ○ CUTE ○ ROUND ○ THIN					
DOES IT MAKE ANY SOUND?	○ YES ○ NO	WAS IT ALONE OR IN A GROUP?		○ ALONE ○ GROUP			

NOTES

PHOTO / DRAWING

BUG IDENTIFICATION

DATE:		TIME:		SEASON:	○ SPRING ○ SUMMER ○ FALL ○ WINTER

WEATHER CONDITIONS:	○ HOT ○ WARM ○ SUNNY ○ CLOUDY ○ RAINY ○ WINDY ○ FOGGY ○ COLD

BUG NAME:	
SCIENTIFIC NAME:	
WHERE DID YOU FIND IT?	
WHAT COLOR(S) IS THE BUG?	

NUMBER OF LEGS?		DOES IT HAVE WINGS?	○ YES ○ NO ○ NOT SURE

THE BUG IS...	○ BIG ○ SHINY ○ FAST ○ SCARY ○ LITTLE ○ SLOW ○ CUTE ○ ROUND ○ THIN

DOES IT MAKE ANY SOUND?	○ YES ○ NO	WAS IT ALONE OR IN A GROUP?	○ ALONE ○ GROUP

NOTES

PHOTO / DRAWING

BUG IDENTIFICATION

DATE:		TIME:		SEASON:	○ SPRING ○ SUMMER ○ FALL ○ WINTER

WEATHER CONDITIONS:	○ HOT ○ WARM ○ SUNNY ○ CLOUDY ○ RAINY ○ WINDY ○ FOGGY ○ COLD
BUG NAME:	
SCIENTIFIC NAME:	
WHERE DID YOU FIND IT?	
WHAT COLOR(S) IS THE BUG?	

NUMBER OF LEGS?		DOES IT HAVE WINGS?	○ YES ○ NO ○ NOT SURE

THE BUG IS...	○ BIG ○ SHINY ○ FAST ○ SCARY ○ LITTLE ○ SLOW ○ CUTE ○ ROUND ○ THIN

DOES IT MAKE ANY SOUND?	○ YES ○ NO	WAS IT ALONE OR IN A GROUP?	○ ALONE ○ GROUP

NOTES

PHOTO / DRAWING

BUG IDENTIFICATION

DATE:		TIME:		SEASON:	○ SPRING ○ SUMMER ○ FALL ○ WINTER

WEATHER CONDITIONS:	○ HOT ○ WARM ○ SUNNY ○ CLOUDY ○ RAINY ○ WINDY ○ FOGGY ○ COLD

BUG NAME:	
SCIENTIFIC NAME:	
WHERE DID YOU FIND IT?	
WHAT COLOR(S) IS THE BUG?	

NUMBER OF LEGS?		DOES IT HAVE WINGS?	○ YES ○ NO ○ NOT SURE

THE BUG IS...	○ BIG ○ SHINY ○ FAST ○ SCARY ○ LITTLE ○ SLOW ○ CUTE ○ ROUND ○ THIN

DOES IT MAKE ANY SOUND?	○ YES ○ NO	WAS IT ALONE OR IN A GROUP?	○ ALONE ○ GROUP

NOTES

PHOTO/DRAWING

BUG IDENTIFICATION

DATE:		TIME:		SEASON:	○ SPRING ○ SUMMER ○ FALL ○ WINTER
WEATHER CONDITIONS:		○ HOT ○ WARM ○ SUNNY ○ CLOUDY ○ RAINY ○ WINDY ○ FOGGY ○ COLD			
BUG NAME:					
SCIENTIFIC NAME:					
WHERE DID YOU FIND IT?					
WHAT COLOR(S) IS THE BUG?					
NUMBER OF LEGS?		**DOES IT HAVE WINGS?**		○ YES ○ NO ○ NOT SURE	
THE BUG IS...		○ BIG ○ SHINY ○ FAST ○ SCARY ○ LITTLE ○ SLOW ○ CUTE ○ ROUND ○ THIN			
DOES IT MAKE ANY SOUND?	○ YES ○ NO	**WAS IT ALONE OR IN A GROUP?**		○ ALONE ○ GROUP	

NOTES

PHOTO/ DRAWING

BUG IDENTIFICATION

DATE:		TIME:		SEASON:	○ SPRING ○ SUMMER ○ FALL ○ WINTER

WEATHER CONDITIONS:	○ HOT ○ WARM ○ SUNNY ○ CLOUDY ○ RAINY ○ WINDY ○ FOGGY ○ COLD
BUG NAME:	
SCIENTIFIC NAME:	
WHERE DID YOU FIND IT?	
WHAT COLOR(S) IS THE BUG?	

NUMBER OF LEGS?		DOES IT HAVE WINGS?	○ YES ○ NO ○ NOT SURE

THE BUG IS...	○ BIG ○ SHINY ○ FAST ○ SCARY ○ LITTLE ○ SLOW ○ CUTE ○ ROUND ○ THIN

DOES IT MAKE ANY SOUND?	○ YES ○ NO	WAS IT ALONE OR IN A GROUP?	○ ALONE ○ GROUP

NOTES

PHOTO/ DRAWING

BUG IDENTIFICATION

DATE:		TIME:		SEASON:	○ SPRING ○ SUMMER ○ FALL ○ WINTER		
WEATHER CONDITIONS:		○ HOT ○ WARM ○ SUNNY ○ CLOUDY ○ RAINY ○ WINDY ○ FOGGY ○ COLD					
BUG NAME:							
SCIENTIFIC NAME:							
WHERE DID YOU FIND IT?							
WHAT COLOR(S) IS THE BUG?							
NUMBER OF LEGS?			**DOES IT HAVE WINGS?**		○ YES ○ NO ○ NOT SURE		
THE BUG IS...		○ BIG ○ SHINY ○ FAST ○ SCARY ○ LITTLE ○ SLOW ○ CUTE ○ ROUND ○ THIN					
DOES IT MAKE ANY SOUND?		○ YES ○ NO	**WAS IT ALONE OR IN A GROUP?**		○ ALONE ○ GROUP		

NOTES

PHOTO / DRAWING

BUG IDENTIFICATION

DATE:		TIME:		SEASON:	○ SPRING ○ SUMMER ○ FALL ○ WINTER

WEATHER CONDITIONS:	○ HOT ○ WARM ○ SUNNY ○ CLOUDY ○ RAINY ○ WINDY ○ FOGGY ○ COLD
BUG NAME:	
SCIENTIFIC NAME:	
WHERE DID YOU FIND IT?	
WHAT COLOR(S) IS THE BUG?	

NUMBER OF LEGS?		DOES IT HAVE WINGS?	○ YES ○ NO ○ NOT SURE

THE BUG IS...	○ BIG ○ SHINY ○ FAST ○ SCARY ○ LITTLE ○ SLOW ○ CUTE ○ ROUND ○ THIN

DOES IT MAKE ANY SOUND?	○ YES ○ NO	WAS IT ALONE OR IN A GROUP?	○ ALONE ○ GROUP

NOTES

PHOTO / DRAWING

BUG IDENTIFICATION

DATE:		TIME:		SEASON:	○ SPRING ○ SUMMER ○ FALL ○ WINTER		
WEATHER CONDITIONS:		○ HOT ○ WARM ○ SUNNY ○ CLOUDY ○ RAINY ○ WINDY ○ FOGGY ○ COLD					
BUG NAME:							
SCIENTIFIC NAME:							
WHERE DID YOU FIND IT?							
WHAT COLOR(S) IS THE BUG?							
NUMBER OF LEGS?		**DOES IT HAVE WINGS?**		○ YES ○ NO ○ NOT SURE			
THE BUG IS...		○ BIG ○ SHINY ○ FAST ○ SCARY ○ LITTLE ○ SLOW ○ CUTE ○ ROUND ○ THIN					
DOES IT MAKE ANY SOUND?	○ YES ○ NO	**WAS IT ALONE OR IN A GROUP?**		○ ALONE ○ GROUP			

NOTES

PHOTO / DRAWING

BUG IDENTIFICATION

DATE:		TIME:		SEASON:	○ SPRING ○ SUMMER ○ FALL ○ WINTER
WEATHER CONDITIONS:		○ HOT ○ WARM ○ SUNNY ○ CLOUDY ○ RAINY ○ WINDY ○ FOGGY ○ COLD			
BUG NAME:					
SCIENTIFIC NAME:					
WHERE DID YOU FIND IT?					
WHAT COLOR(S) IS THE BUG?					
NUMBER OF LEGS?		**DOES IT HAVE WINGS?**		○ YES ○ NO ○ NOT SURE	
THE BUG IS...		○ BIG ○ SHINY ○ FAST ○ SCARY ○ LITTLE ○ SLOW ○ CUTE ○ ROUND ○ THIN			
DOES IT MAKE ANY SOUND?	○ YES ○ NO	**WAS IT ALONE OR IN A GROUP?**		○ ALONE ○ GROUP	

NOTES

PHOTO/DRAWING

BUG IDENTIFICATION

DATE:		TIME:		SEASON:	○ SPRING ○ SUMMER ○ FALL ○ WINTER		
WEATHER CONDITIONS:		○ HOT ○ WARM ○ SUNNY ○ CLOUDY ○ RAINY ○ WINDY ○ FOGGY ○ COLD					
BUG NAME:							
SCIENTIFIC NAME:							
WHERE DID YOU FIND IT?							
WHAT COLOR(S) IS THE BUG?							
NUMBER OF LEGS?			DOES IT HAVE WINGS?		○ YES ○ NO ○ NOT SURE		
THE BUG IS...		○ BIG ○ SHINY ○ FAST ○ SCARY ○ LITTLE ○ SLOW ○ CUTE ○ ROUND ○ THIN					
DOES IT MAKE ANY SOUND?		○ YES ○ NO	WAS IT ALONE OR IN A GROUP?		○ ALONE ○ GROUP		

NOTES

PHOTO / DRAWING

BUG IDENTIFICATION

DATE:		TIME:		SEASON:	○ SPRING ○ SUMMER ○ FALL ○ WINTER		
WEATHER CONDITIONS:		○ HOT ○ WARM ○ SUNNY ○ CLOUDY ○ RAINY ○ WINDY ○ FOGGY ○ COLD					
BUG NAME:							
SCIENTIFIC NAME:							
WHERE DID YOU FIND IT?							
WHAT COLOR(S) IS THE BUG?							
NUMBER OF LEGS?		DOES IT HAVE WINGS?		○ YES ○ NO ○ NOT SURE			
THE BUG IS...		○ BIG ○ SHINY ○ FAST ○ SCARY ○ LITTLE ○ SLOW ○ CUTE ○ ROUND ○ THIN					
DOES IT MAKE ANY SOUND?		○ YES ○ NO	WAS IT ALONE OR IN A GROUP?		○ ALONE ○ GROUP		

NOTES

PHOTO/ DRAWING

BUG IDENTIFICATION

DATE:		TIME:		SEASON:	○ SPRING ○ SUMMER ○ FALL ○ WINTER		
WEATHER CONDITIONS:		○ HOT ○ WARM ○ SUNNY ○ CLOUDY ○ RAINY ○ WINDY ○ FOGGY ○ COLD					
BUG NAME:							
SCIENTIFIC NAME:							
WHERE DID YOU FIND IT?							
WHAT COLOR(S) IS THE BUG?							
NUMBER OF LEGS?		**DOES IT HAVE WINGS?**		○ YES ○ NO ○ NOT SURE			
THE BUG IS...		○ BIG ○ SHINY ○ FAST ○ SCARY ○ LITTLE ○ SLOW ○ CUTE ○ ROUND ○ THIN					
DOES IT MAKE ANY SOUND?		○ YES ○ NO	**WAS IT ALONE OR IN A GROUP?**		○ ALONE ○ GROUP		

NOTES

PHOTO / DRAWING

BUG IDENTIFICATION

DATE:		TIME:		SEASON:	○ SPRING ○ SUMMER ○ FALL ○ WINTER

WEATHER CONDITIONS:	○ HOT ○ WARM ○ SUNNY ○ CLOUDY ○ RAINY ○ WINDY ○ FOGGY ○ COLD

BUG NAME:	
SCIENTIFIC NAME:	
WHERE DID YOU FIND IT?	
WHAT COLOR(S) IS THE BUG?	

NUMBER OF LEGS?		DOES IT HAVE WINGS?	○ YES ○ NO ○ NOT SURE

THE BUG IS...	○ BIG ○ SHINY ○ FAST ○ SCARY ○ LITTLE ○ SLOW ○ CUTE ○ ROUND ○ THIN

DOES IT MAKE ANY SOUND?	○ YES ○ NO	WAS IT ALONE OR IN A GROUP?	○ ALONE ○ GROUP

NOTES

PHOTO/DRAWING

BUG IDENTIFICATION

DATE:		TIME:		SEASON:	○ SPRING ○ SUMMER ○ FALL ○ WINTER

WEATHER CONDITIONS:	○ HOT ○ WARM ○ SUNNY ○ CLOUDY ○ RAINY ○ WINDY ○ FOGGY ○ COLD
BUG NAME:	
SCIENTIFIC NAME:	
WHERE DID YOU FIND IT?	
WHAT COLOR(S) IS THE BUG?	

NUMBER OF LEGS?		DOES IT HAVE WINGS?	○ YES ○ NO ○ NOT SURE
THE BUG IS...	○ BIG ○ SHINY ○ FAST ○ SCARY ○ LITTLE ○ SLOW ○ CUTE ○ ROUND ○ THIN		
DOES IT MAKE ANY SOUND?	○ YES ○ NO	WAS IT ALONE OR IN A GROUP?	○ ALONE ○ GROUP

NOTES

PHOTO / DRAWING

BUG IDENTIFICATION

DATE:		TIME:		SEASON:	○ SPRING ○ SUMMER ○ FALL ○ WINTER		
WEATHER CONDITIONS:		○ HOT ○ WARM ○ SUNNY ○ CLOUDY ○ RAINY ○ WINDY ○ FOGGY ○ COLD					
BUG NAME:							
SCIENTIFIC NAME:							
WHERE DID YOU FIND IT?							
WHAT COLOR(S) IS THE BUG?							
NUMBER OF LEGS?		DOES IT HAVE WINGS?		○ YES ○ NO ○ NOT SURE			
THE BUG IS...		○ BIG ○ SHINY ○ FAST ○ SCARY ○ LITTLE ○ SLOW ○ CUTE ○ ROUND ○ THIN					
DOES IT MAKE ANY SOUND?		○ YES ○ NO	WAS IT ALONE OR IN A GROUP?		○ ALONE ○ GROUP		

NOTES

PHOTO/DRAWING

BUG IDENTIFICATION

DATE:		TIME:		SEASON:	○ SPRING ○ SUMMER ○ FALL ○ WINTER		
WEATHER CONDITIONS:		○ HOT ○ WARM ○ SUNNY ○ CLOUDY ○ RAINY ○ WINDY ○ FOGGY ○ COLD					
BUG NAME:							
SCIENTIFIC NAME:							
WHERE DID YOU FIND IT?							
WHAT COLOR(S) IS THE BUG?							
NUMBER OF LEGS?			DOES IT HAVE WINGS?		○ YES ○ NO ○ NOT SURE		
THE BUG IS...		○ BIG ○ SHINY ○ FAST ○ SCARY ○ LITTLE ○ SLOW ○ CUTE ○ ROUND ○ THIN					
DOES IT MAKE ANY SOUND?	○ YES ○ NO		WAS IT ALONE OR IN A GROUP?		○ ALONE ○ GROUP		

NOTES

PHOTO/ DRAWING

BUG IDENTIFICATION

DATE:		TIME:		SEASON:	○ SPRING ○ SUMMER ○ FALL ○ WINTER

WEATHER CONDITIONS:	○ HOT ○ WARM ○ SUNNY ○ CLOUDY ○ RAINY ○ WINDY ○ FOGGY ○ COLD

BUG NAME:	
SCIENTIFIC NAME:	
WHERE DID YOU FIND IT?	
WHAT COLOR(S) IS THE BUG?	

NUMBER OF LEGS?		DOES IT HAVE WINGS?	○ YES ○ NO ○ NOT SURE

THE BUG IS...	○ BIG ○ SHINY ○ FAST ○ SCARY ○ LITTLE ○ SLOW ○ CUTE ○ ROUND ○ THIN

DOES IT MAKE ANY SOUND?	○ YES ○ NO	WAS IT ALONE OR IN A GROUP?	○ ALONE ○ GROUP

NOTES

PHOTO / DRAWING

BUG IDENTIFICATION

DATE:		TIME:		SEASON:	○ SPRING ○ SUMMER ○ FALL ○ WINTER			
WEATHER CONDITIONS:		○ HOT ○ WARM ○ SUNNY ○ CLOUDY ○ RAINY ○ WINDY ○ FOGGY ○ COLD						
BUG NAME:								
SCIENTIFIC NAME:								
WHERE DID YOU FIND IT?								
WHAT COLOR(S) IS THE BUG?								
NUMBER OF LEGS?			**DOES IT HAVE WINGS?**		○ YES ○ NO ○ NOT SURE			
THE BUG IS...		○ BIG ○ SHINY ○ FAST ○ SCARY ○ LITTLE ○ SLOW ○ CUTE ○ ROUND ○ THIN						
DOES IT MAKE ANY SOUND?	○ YES ○ NO		**WAS IT ALONE OR IN A GROUP?**		○ ALONE ○ GROUP			

NOTES

PHOTO / DRAWING

BUG IDENTIFICATION

DATE:		TIME:		SEASON:	○ SPRING ○ SUMMER ○ FALL ○ WINTER
WEATHER CONDITIONS:		○ HOT ○ WARM ○ SUNNY ○ CLOUDY ○ RAINY ○ WINDY ○ FOGGY ○ COLD			
BUG NAME:					
SCIENTIFIC NAME:					
WHERE DID YOU FIND IT?					
WHAT COLOR(S) IS THE BUG?					
NUMBER OF LEGS?		**DOES IT HAVE WINGS?**		○ YES ○ NO ○ NOT SURE	
THE BUG IS...		○ BIG ○ SHINY ○ FAST ○ SCARY ○ LITTLE ○ SLOW ○ CUTE ○ ROUND ○ THIN			
DOES IT MAKE ANY SOUND?	○ YES ○ NO	**WAS IT ALONE OR IN A GROUP?**		○ ALONE ○ GROUP	

NOTES

PHOTO / DRAWING

BUG IDENTIFICATION

DATE:		TIME:		SEASON:	○ SPRING ○ SUMMER ○ FALL ○ WINTER

WEATHER CONDITIONS:	○ HOT ○ WARM ○ SUNNY ○ CLOUDY ○ RAINY ○ WINDY ○ FOGGY ○ COLD
BUG NAME:	
SCIENTIFIC NAME:	
WHERE DID YOU FIND IT?	
WHAT COLOR(S) IS THE BUG?	

NUMBER OF LEGS?		DOES IT HAVE WINGS?		○ YES ○ NO ○ NOT SURE

THE BUG IS...	○ BIG ○ SHINY ○ FAST ○ SCARY ○ LITTLE ○ SLOW ○ CUTE ○ ROUND ○ THIN

DOES IT MAKE ANY SOUND?	○ YES ○ NO	WAS IT ALONE OR IN A GROUP?	○ ALONE ○ GROUP

NOTES

PHOTO / DRAWING

BUG IDENTIFICATION

DATE:		TIME:		SEASON:	○ SPRING ○ SUMMER ○ FALL ○ WINTER
WEATHER CONDITIONS:		○ HOT ○ WARM ○ SUNNY ○ CLOUDY ○ RAINY ○ WINDY ○ FOGGY ○ COLD			
BUG NAME:					
SCIENTIFIC NAME:					
WHERE DID YOU FIND IT?					
WHAT COLOR(S) IS THE BUG?					
NUMBER OF LEGS?		**DOES IT HAVE WINGS?**		○ YES ○ NO ○ NOT SURE	
THE BUG IS...		○ BIG ○ SHINY ○ FAST ○ SCARY ○ LITTLE ○ SLOW ○ CUTE ○ ROUND ○ THIN			
DOES IT MAKE ANY SOUND?	○ YES ○ NO	**WAS IT ALONE OR IN A GROUP?**		○ ALONE ○ GROUP	

NOTES

PHOTO/ DRAWING

BUG IDENTIFICATION

DATE:		TIME:		SEASON:	○ SPRING ○ SUMMER ○ FALL ○ WINTER		
WEATHER CONDITIONS:		○ HOT ○ WARM ○ SUNNY ○ CLOUDY ○ RAINY ○ WINDY ○ FOGGY ○ COLD					
BUG NAME:							
SCIENTIFIC NAME:							
WHERE DID YOU FIND IT?							
WHAT COLOR(S) IS THE BUG?							
NUMBER OF LEGS?			DOES IT HAVE WINGS?		○ YES ○ NO ○ NOT SURE		
THE BUG IS...		○ BIG ○ SHINY ○ FAST ○ SCARY ○ LITTLE ○ SLOW ○ CUTE ○ ROUND ○ THIN					
DOES IT MAKE ANY SOUND?		○ YES ○ NO	WAS IT ALONE OR IN A GROUP?		○ ALONE ○ GROUP		

NOTES

PHOTO / DRAWING

BUG IDENTIFICATION

DATE:		TIME:		SEASON:	○ SPRING ○ SUMMER ○ FALL ○ WINTER

WEATHER CONDITIONS:	○ HOT ○ WARM ○ SUNNY ○ CLOUDY ○ RAINY ○ WINDY ○ FOGGY ○ COLD

BUG NAME:	
SCIENTIFIC NAME:	
WHERE DID YOU FIND IT?	
WHAT COLOR(S) IS THE BUG?	

NUMBER OF LEGS?		DOES IT HAVE WINGS?	○ YES ○ NO ○ NOT SURE

THE BUG IS...	○ BIG ○ SHINY ○ FAST ○ SCARY ○ LITTLE ○ SLOW ○ CUTE ○ ROUND ○ THIN

DOES IT MAKE ANY SOUND?	○ YES ○ NO	WAS IT ALONE OR IN A GROUP?	○ ALONE ○ GROUP

NOTES

PHOTO/DRAWING

BUG IDENTIFICATION

DATE:		TIME:		SEASON:	○ SPRING ○ SUMMER ○ FALL ○ WINTER

WEATHER CONDITIONS:	○ HOT ○ WARM ○ SUNNY ○ CLOUDY ○ RAINY ○ WINDY ○ FOGGY ○ COLD
BUG NAME:	
SCIENTIFIC NAME:	
WHERE DID YOU FIND IT?	
WHAT COLOR(S) IS THE BUG?	

NUMBER OF LEGS?		DOES IT HAVE WINGS?		○ YES ○ NO ○ NOT SURE

THE BUG IS...	○ BIG ○ SHINY ○ FAST ○ SCARY ○ LITTLE ○ SLOW ○ CUTE ○ ROUND ○ THIN

DOES IT MAKE ANY SOUND?	○ YES ○ NO	WAS IT ALONE OR IN A GROUP?	○ ALONE ○ GROUP

NOTES

PHOTO / DRAWING

BUG IDENTIFICATION

DATE:		TIME:		SEASON:	○ SPRING ○ SUMMER ○ FALL ○ WINTER		
WEATHER CONDITIONS:		○ HOT ○ WARM ○ SUNNY ○ CLOUDY ○ RAINY ○ WINDY ○ FOGGY ○ COLD					
BUG NAME:							
SCIENTIFIC NAME:							
WHERE DID YOU FIND IT?							
WHAT COLOR(S) IS THE BUG?							
NUMBER OF LEGS?		**DOES IT HAVE WINGS?**		○ YES ○ NO ○ NOT SURE			
THE BUG IS...		○ BIG ○ SHINY ○ FAST ○ SCARY ○ LITTLE ○ SLOW ○ CUTE ○ ROUND ○ THIN					
DOES IT MAKE ANY SOUND?		○ YES ○ NO	**WAS IT ALONE OR IN A GROUP?**		○ ALONE ○ GROUP		

NOTES

PHOTO / DRAWING

BUG IDENTIFICATION

DATE:		TIME:		SEASON:	○ SPRING ○ SUMMER ○ FALL ○ WINTER

WEATHER CONDITIONS:	○ HOT ○ WARM ○ SUNNY ○ CLOUDY ○ RAINY ○ WINDY ○ FOGGY ○ COLD

BUG NAME:	
SCIENTIFIC NAME:	
WHERE DID YOU FIND IT?	
WHAT COLOR(S) IS THE BUG?	

NUMBER OF LEGS?		DOES IT HAVE WINGS?	○ YES ○ NO ○ NOT SURE

THE BUG IS...	○ BIG ○ SHINY ○ FAST ○ SCARY ○ LITTLE ○ SLOW ○ CUTE ○ ROUND ○ THIN

DOES IT MAKE ANY SOUND?	○ YES ○ NO	WAS IT ALONE OR IN A GROUP?	○ ALONE ○ GROUP

NOTES

PHOTO/ DRAWING

BUG IDENTIFICATION

DATE:		TIME:		SEASON:	○ SPRING ○ SUMMER ○ FALL ○ WINTER

WEATHER CONDITIONS:	○ HOT ○ WARM ○ SUNNY ○ CLOUDY ○ RAINY ○ WINDY ○ FOGGY ○ COLD
BUG NAME:	
SCIENTIFIC NAME:	
WHERE DID YOU FIND IT?	
WHAT COLOR(S) IS THE BUG?	

NUMBER OF LEGS?		DOES IT HAVE WINGS?	○ YES ○ NO ○ NOT SURE

THE BUG IS...	○ BIG ○ SHINY ○ FAST ○ SCARY ○ LITTLE ○ SLOW ○ CUTE ○ ROUND ○ THIN

DOES IT MAKE ANY SOUND?	○ YES ○ NO	WAS IT ALONE OR IN A GROUP?	○ ALONE ○ GROUP

NOTES

PHOTO/DRAWING

BUG IDENTIFICATION

DATE:		TIME:		SEASON:	○ SPRING ○ SUMMER ○ FALL ○ WINTER

WEATHER CONDITIONS:	○ HOT ○ WARM ○ SUNNY ○ CLOUDY ○ RAINY ○ WINDY ○ FOGGY ○ COLD

BUG NAME:	
SCIENTIFIC NAME:	
WHERE DID YOU FIND IT?	
WHAT COLOR(S) IS THE BUG?	

NUMBER OF LEGS?		DOES IT HAVE WINGS?	○ YES ○ NO ○ NOT SURE

THE BUG IS...	○ BIG ○ SHINY ○ FAST ○ SCARY ○ LITTLE ○ SLOW ○ CUTE ○ ROUND ○ THIN

DOES IT MAKE ANY SOUND?	○ YES ○ NO	WAS IT ALONE OR IN A GROUP?	○ ALONE ○ GROUP

NOTES

PHOTO / DRAWING

BUG IDENTIFICATION

DATE:		TIME:		SEASON:	○ SPRING ○ SUMMER ○ FALL ○ WINTER

WEATHER CONDITIONS:	○ HOT ○ WARM ○ SUNNY ○ CLOUDY ○ RAINY ○ WINDY ○ FOGGY ○ COLD
BUG NAME:	
SCIENTIFIC NAME:	
WHERE DID YOU FIND IT?	
WHAT COLOR(S) IS THE BUG?	

NUMBER OF LEGS?		DOES IT HAVE WINGS?	○ YES ○ NO ○ NOT SURE

THE BUG IS...	○ BIG ○ SHINY ○ FAST ○ SCARY ○ LITTLE ○ SLOW ○ CUTE ○ ROUND ○ THIN

DOES IT MAKE ANY SOUND?	○ YES ○ NO	WAS IT ALONE OR IN A GROUP?	○ ALONE ○ GROUP

NOTES

PHOTO / DRAWING

BUG IDENTIFICATION

DATE:		TIME:		SEASON:	○ SPRING ○ SUMMER ○ FALL ○ WINTER		
WEATHER CONDITIONS:		○ HOT ○ WARM ○ SUNNY ○ CLOUDY ○ RAINY ○ WINDY ○ FOGGY ○ COLD					
BUG NAME:							
SCIENTIFIC NAME:							
WHERE DID YOU FIND IT?							
WHAT COLOR(S) IS THE BUG?							
NUMBER OF LEGS?		**DOES IT HAVE WINGS?**		○ YES ○ NO ○ NOT SURE			
THE BUG IS...		○ BIG ○ SHINY ○ FAST ○ SCARY ○ LITTLE ○ SLOW ○ CUTE ○ ROUND ○ THIN					
DOES IT MAKE ANY SOUND?	○ YES ○ NO	**WAS IT ALONE OR IN A GROUP?**		○ ALONE ○ GROUP			

NOTES

PHOTO/DRAWING

BUG IDENTIFICATION

DATE:		TIME:		SEASON:	○ SPRING ○ SUMMER ○ FALL ○ WINTER

WEATHER CONDITIONS:	○ HOT ○ WARM ○ SUNNY ○ CLOUDY ○ RAINY ○ WINDY ○ FOGGY ○ COLD
BUG NAME:	
SCIENTIFIC NAME:	
WHERE DID YOU FIND IT?	
WHAT COLOR(S) IS THE BUG?	

NUMBER OF LEGS?		DOES IT HAVE WINGS?		○ YES ○ NO ○ NOT SURE

THE BUG IS...	○ BIG ○ SHINY ○ FAST ○ SCARY ○ LITTLE ○ SLOW ○ CUTE ○ ROUND ○ THIN

DOES IT MAKE ANY SOUND?	○ YES ○ NO	WAS IT ALONE OR IN A GROUP?	○ ALONE ○ GROUP

NOTES

PHOTO / DRAWING

BUG IDENTIFICATION

DATE:		TIME:		SEASON:	○ SPRING ○ SUMMER ○ FALL ○ WINTER

WEATHER CONDITIONS:	○ HOT ○ WARM ○ SUNNY ○ CLOUDY ○ RAINY ○ WINDY ○ FOGGY ○ COLD

BUG NAME:	
SCIENTIFIC NAME:	
WHERE DID YOU FIND IT?	
WHAT COLOR(S) IS THE BUG?	

NUMBER OF LEGS?		DOES IT HAVE WINGS?	○ YES ○ NO ○ NOT SURE

THE BUG IS...	○ BIG ○ SHINY ○ FAST ○ SCARY ○ LITTLE ○ SLOW ○ CUTE ○ ROUND ○ THIN

DOES IT MAKE ANY SOUND?	○ YES ○ NO	WAS IT ALONE OR IN A GROUP?	○ ALONE ○ GROUP

NOTES

PHOTO/DRAWING

BUG IDENTIFICATION

DATE:		TIME:		SEASON:	○ SPRING ○ SUMMER ○ FALL ○ WINTER

WEATHER CONDITIONS:	○ HOT ○ WARM ○ SUNNY ○ CLOUDY ○ RAINY ○ WINDY ○ FOGGY ○ COLD
BUG NAME:	
SCIENTIFIC NAME:	
WHERE DID YOU FIND IT?	
WHAT COLOR(S) IS THE BUG?	

NUMBER OF LEGS?		DOES IT HAVE WINGS?		○ YES ○ NO ○ NOT SURE

THE BUG IS...	○ BIG ○ SHINY ○ FAST ○ SCARY ○ LITTLE ○ SLOW ○ CUTE ○ ROUND ○ THIN

DOES IT MAKE ANY SOUND?	○ YES ○ NO	WAS IT ALONE OR IN A GROUP?	○ ALONE ○ GROUP

NOTES

PHOTO/DRAWING

BUG IDENTIFICATION

DATE:		TIME:		SEASON:	○ SPRING ○ SUMMER ○ FALL ○ WINTER

WEATHER CONDITIONS:	○ HOT ○ WARM ○ SUNNY ○ CLOUDY ○ RAINY ○ WINDY ○ FOGGY ○ COLD

BUG NAME:	
SCIENTIFIC NAME:	
WHERE DID YOU FIND IT?	
WHAT COLOR(S) IS THE BUG?	

NUMBER OF LEGS?		DOES IT HAVE WINGS?	○ YES ○ NO ○ NOT SURE

THE BUG IS...	○ BIG ○ SHINY ○ FAST ○ SCARY ○ LITTLE ○ SLOW ○ CUTE ○ ROUND ○ THIN

DOES IT MAKE ANY SOUND?	○ YES ○ NO	WAS IT ALONE OR IN A GROUP?	○ ALONE ○ GROUP

NOTES

PHOTO / DRAWING

BUG IDENTIFICATION

DATE:		TIME:		SEASON:	○ SPRING ○ SUMMER ○ FALL ○ WINTER

WEATHER CONDITIONS:	○ HOT ○ WARM ○ SUNNY ○ CLOUDY ○ RAINY ○ WINDY ○ FOGGY ○ COLD
BUG NAME:	
SCIENTIFIC NAME:	
WHERE DID YOU FIND IT?	
WHAT COLOR(S) IS THE BUG?	

NUMBER OF LEGS?		DOES IT HAVE WINGS?	○ YES ○ NO ○ NOT SURE

THE BUG IS...	○ BIG ○ SHINY ○ FAST ○ SCARY ○ LITTLE ○ SLOW ○ CUTE ○ ROUND ○ THIN

DOES IT MAKE ANY SOUND?	○ YES ○ NO	WAS IT ALONE OR IN A GROUP?	○ ALONE ○ GROUP

NOTES

PHOTO/DRAWING

BUG IDENTIFICATION

DATE:		TIME:		SEASON:	○ SPRING ○ SUMMER ○ FALL ○ WINTER

WEATHER CONDITIONS:	○ HOT ○ WARM ○ SUNNY ○ CLOUDY ○ RAINY ○ WINDY ○ FOGGY ○ COLD
BUG NAME:	
SCIENTIFIC NAME:	
WHERE DID YOU FIND IT?	
WHAT COLOR(S) IS THE BUG?	

NUMBER OF LEGS?		DOES IT HAVE WINGS?	○ YES ○ NO ○ NOT SURE

THE BUG IS...	○ BIG ○ SHINY ○ FAST ○ SCARY ○ LITTLE ○ SLOW ○ CUTE ○ ROUND ○ THIN

DOES IT MAKE ANY SOUND?	○ YES ○ NO	WAS IT ALONE OR IN A GROUP?	○ ALONE ○ GROUP

NOTES

PHOTO / DRAWING

BUG IDENTIFICATION

DATE:		TIME:		SEASON:	○ SPRING ○ SUMMER ○ FALL ○ WINTER		
WEATHER CONDITIONS:		○ HOT ○ WARM ○ SUNNY ○ CLOUDY ○ RAINY ○ WINDY ○ FOGGY ○ COLD					
BUG NAME:							
SCIENTIFIC NAME:							
WHERE DID YOU FIND IT?							
WHAT COLOR(S) IS THE BUG?							
NUMBER OF LEGS?			DOES IT HAVE WINGS?		○ YES ○ NO ○ NOT SURE		
THE BUG IS...		○ BIG ○ SHINY ○ FAST ○ SCARY ○ LITTLE ○ SLOW ○ CUTE ○ ROUND ○ THIN					
DOES IT MAKE ANY SOUND?		○ YES ○ NO	WAS IT ALONE OR IN A GROUP?			○ ALONE ○ GROUP	

NOTES

PHOTO / DRAWING

BUG IDENTIFICATION

DATE:		TIME:		SEASON:	○ SPRING ○ SUMMER ○ FALL ○ WINTER

WEATHER CONDITIONS:	○ HOT ○ WARM ○ SUNNY ○ CLOUDY ○ RAINY ○ WINDY ○ FOGGY ○ COLD
BUG NAME:	
SCIENTIFIC NAME:	
WHERE DID YOU FIND IT?	
WHAT COLOR(S) IS THE BUG?	

NUMBER OF LEGS?		DOES IT HAVE WINGS?	○ YES ○ NO ○ NOT SURE

THE BUG IS...	○ BIG ○ SHINY ○ FAST ○ SCARY ○ LITTLE ○ SLOW ○ CUTE ○ ROUND ○ THIN

DOES IT MAKE ANY SOUND?	○ YES ○ NO	WAS IT ALONE OR IN A GROUP?	○ ALONE ○ GROUP

NOTES

PHOTO/DRAWING

BUG IDENTIFICATION

DATE:		TIME:		SEASON:	○ SPRING ○ SUMMER ○ FALL ○ WINTER
WEATHER CONDITIONS:			○ HOT ○ WARM ○ SUNNY ○ CLOUDY ○ RAINY ○ WINDY ○ FOGGY ○ COLD		
BUG NAME:					
SCIENTIFIC NAME:					
WHERE DID YOU FIND IT?					
WHAT COLOR(S) IS THE BUG?					
NUMBER OF LEGS?		DOES IT HAVE WINGS?		○ YES ○ NO ○ NOT SURE	
THE BUG IS...	○ BIG ○ SHINY ○ FAST ○ SCARY ○ LITTLE ○ SLOW ○ CUTE ○ ROUND ○ THIN				
DOES IT MAKE ANY SOUND?	○ YES ○ NO	WAS IT ALONE OR IN A GROUP?		○ ALONE ○ GROUP	

NOTES

PHOTO/DRAWING

BUG IDENTIFICATION

DATE:		TIME:		SEASON:	○ SPRING ○ SUMMER ○ FALL ○ WINTER		
WEATHER CONDITIONS:		○ HOT ○ WARM ○ SUNNY ○ CLOUDY ○ RAINY ○ WINDY ○ FOGGY ○ COLD					
BUG NAME:							
SCIENTIFIC NAME:							
WHERE DID YOU FIND IT?							
WHAT COLOR(S) IS THE BUG?							
NUMBER OF LEGS?			DOES IT HAVE WINGS?		○ YES ○ NO ○ NOT SURE		
THE BUG IS...		○ BIG ○ SHINY ○ FAST ○ SCARY ○ LITTLE ○ SLOW ○ CUTE ○ ROUND ○ THIN					
DOES IT MAKE ANY SOUND?		○ YES ○ NO	WAS IT ALONE OR IN A GROUP?		○ ALONE ○ GROUP		

NOTES

PHOTO / DRAWING

BUG IDENTIFICATION

DATE:		TIME:		SEASON:	○ SPRING ○ SUMMER ○ FALL ○ WINTER

WEATHER CONDITIONS:	○ HOT ○ WARM ○ SUNNY ○ CLOUDY ○ RAINY ○ WINDY ○ FOGGY ○ COLD

BUG NAME:	
SCIENTIFIC NAME:	
WHERE DID YOU FIND IT?	
WHAT COLOR(S) IS THE BUG?	

NUMBER OF LEGS?		DOES IT HAVE WINGS?	○ YES ○ NO ○ NOT SURE

THE BUG IS...	○ BIG ○ SHINY ○ FAST ○ SCARY ○ LITTLE ○ SLOW ○ CUTE ○ ROUND ○ THIN

DOES IT MAKE ANY SOUND?	○ YES ○ NO	WAS IT ALONE OR IN A GROUP?	○ ALONE ○ GROUP

NOTES

PHOTO/ DRAWING

BUG IDENTIFICATION

DATE:		TIME:		SEASON:	○ SPRING ○ SUMMER ○ FALL ○ WINTER

WEATHER CONDITIONS:	○ HOT ○ WARM ○ SUNNY ○ CLOUDY ○ RAINY ○ WINDY ○ FOGGY ○ COLD
BUG NAME:	
SCIENTIFIC NAME:	
WHERE DID YOU FIND IT?	
WHAT COLOR(S) IS THE BUG?	

NUMBER OF LEGS?		DOES IT HAVE WINGS?	○ YES ○ NO ○ NOT SURE

THE BUG IS...	○ BIG ○ SHINY ○ FAST ○ SCARY ○ LITTLE ○ SLOW ○ CUTE ○ ROUND ○ THIN

DOES IT MAKE ANY SOUND?	○ YES ○ NO	WAS IT ALONE OR IN A GROUP?	○ ALONE ○ GROUP

NOTES

PHOTO / DRAWING

BUG IDENTIFICATION

DATE:		TIME:		SEASON:	○ SPRING ○ SUMMER ○ FALL ○ WINTER	
WEATHER CONDITIONS:		○ HOT ○ WARM ○ SUNNY ○ CLOUDY ○ RAINY ○ WINDY ○ FOGGY ○ COLD				
BUG NAME:						
SCIENTIFIC NAME:						
WHERE DID YOU FIND IT?						
WHAT COLOR(S) IS THE BUG?						
NUMBER OF LEGS?		**DOES IT HAVE WINGS?**		○ YES ○ NO ○ NOT SURE		
THE BUG IS...		○ BIG ○ SHINY ○ FAST ○ SCARY ○ LITTLE ○ SLOW ○ CUTE ○ ROUND ○ THIN				
DOES IT MAKE ANY SOUND?	○ YES ○ NO	**WAS IT ALONE OR IN A GROUP?**		○ ALONE ○ GROUP		

NOTES

PHOTO / DRAWING

BUG IDENTIFICATION

DATE:		TIME:		SEASON:	○ SPRING ○ SUMMER ○ FALL ○ WINTER

WEATHER CONDITIONS:	○ HOT ○ WARM ○ SUNNY ○ CLOUDY ○ RAINY ○ WINDY ○ FOGGY ○ COLD
BUG NAME:	
SCIENTIFIC NAME:	
WHERE DID YOU FIND IT?	
WHAT COLOR(S) IS THE BUG?	

NUMBER OF LEGS?		DOES IT HAVE WINGS?	○ YES ○ NO ○ NOT SURE

THE BUG IS...	○ BIG ○ SHINY ○ FAST ○ SCARY ○ LITTLE ○ SLOW ○ CUTE ○ ROUND ○ THIN

DOES IT MAKE ANY SOUND?	○ YES ○ NO	WAS IT ALONE OR IN A GROUP?	○ ALONE ○ GROUP

NOTES

PHOTO / DRAWING

BUG IDENTIFICATION

DATE:		TIME:		SEASON:	○ SPRING ○ SUMMER ○ FALL ○ WINTER

WEATHER CONDITIONS:	○ HOT ○ WARM ○ SUNNY ○ CLOUDY ○ RAINY ○ WINDY ○ FOGGY ○ COLD

BUG NAME:	
SCIENTIFIC NAME:	
WHERE DID YOU FIND IT?	
WHAT COLOR(S) IS THE BUG?	

NUMBER OF LEGS?		DOES IT HAVE WINGS?		○ YES ○ NO ○ NOT SURE

THE BUG IS...	○ BIG ○ SHINY ○ FAST ○ SCARY ○ LITTLE ○ SLOW ○ CUTE ○ ROUND ○ THIN

DOES IT MAKE ANY SOUND?	○ YES ○ NO	WAS IT ALONE OR IN A GROUP?	○ ALONE ○ GROUP

NOTES

PHOTO/DRAWING

BUG IDENTIFICATION

DATE:		TIME:		SEASON:	○ SPRING ○ SUMMER ○ FALL ○ WINTER		
WEATHER CONDITIONS:		○ HOT ○ WARM ○ SUNNY ○ CLOUDY ○ RAINY ○ WINDY ○ FOGGY ○ COLD					
BUG NAME:							
SCIENTIFIC NAME:							
WHERE DID YOU FIND IT?							
WHAT COLOR(S) IS THE BUG?							
NUMBER OF LEGS?			**DOES IT HAVE WINGS?**		○ YES ○ NO ○ NOT SURE		
THE BUG IS...		○ BIG ○ SHINY ○ FAST ○ SCARY ○ LITTLE ○ SLOW ○ CUTE ○ ROUND ○ THIN					
DOES IT MAKE ANY SOUND?		○ YES ○ NO	**WAS IT ALONE OR IN A GROUP?**			○ ALONE ○ GROUP	

NOTES

PHOTO / DRAWING

BUG IDENTIFICATION

DATE:		TIME:		SEASON:	○ SPRING ○ SUMMER ○ FALL ○ WINTER		
WEATHER CONDITIONS:		○ HOT ○ WARM ○ SUNNY ○ CLOUDY ○ RAINY ○ WINDY ○ FOGGY ○ COLD					
BUG NAME:							
SCIENTIFIC NAME:							
WHERE DID YOU FIND IT?							
WHAT COLOR(S) IS THE BUG?							
NUMBER OF LEGS?		**DOES IT HAVE WINGS?**		○ YES ○ NO ○ NOT SURE			
THE BUG IS...		○ BIG ○ SHINY ○ FAST ○ SCARY ○ LITTLE ○ SLOW ○ CUTE ○ ROUND ○ THIN					
DOES IT MAKE ANY SOUND?		○ YES ○ NO	**WAS IT ALONE OR IN A GROUP?**		○ ALONE ○ GROUP		

NOTES

PHOTO / DRAWING

BUG IDENTIFICATION

DATE:		TIME:		SEASON:	○ SPRING ○ SUMMER ○ FALL ○ WINTER
WEATHER CONDITIONS:		○ HOT ○ WARM ○ SUNNY ○ CLOUDY ○ RAINY ○ WINDY ○ FOGGY ○ COLD			
BUG NAME:					
SCIENTIFIC NAME:					
WHERE DID YOU FIND IT?					
WHAT COLOR(S) IS THE BUG?					
NUMBER OF LEGS?		**DOES IT HAVE WINGS?**	○ YES ○ NO ○ NOT SURE		
THE BUG IS...		○ BIG ○ SHINY ○ FAST ○ SCARY ○ LITTLE ○ SLOW ○ CUTE ○ ROUND ○ THIN			
DOES IT MAKE ANY SOUND?	○ YES ○ NO	**WAS IT ALONE OR IN A GROUP?**	○ ALONE ○ GROUP		

NOTES

PHOTO / DRAWING

BUG IDENTIFICATION

DATE:		TIME:		SEASON:	○ SPRING ○ SUMMER ○ FALL ○ WINTER		
WEATHER CONDITIONS:		○ HOT ○ WARM ○ SUNNY ○ CLOUDY ○ RAINY ○ WINDY ○ FOGGY ○ COLD					
BUG NAME:							
SCIENTIFIC NAME:							
WHERE DID YOU FIND IT?							
WHAT COLOR(S) IS THE BUG?							
NUMBER OF LEGS?			**DOES IT HAVE WINGS?**	○ YES ○ NO ○ NOT SURE			
THE BUG IS...		○ BIG ○ SHINY ○ FAST ○ SCARY ○ LITTLE ○ SLOW ○ CUTE ○ ROUND ○ THIN					
DOES IT MAKE ANY SOUND?		○ YES ○ NO	**WAS IT ALONE OR IN A GROUP?**	○ ALONE ○ GROUP			

NOTES

PHOTO / DRAWING

BUG IDENTIFICATION

DATE:		TIME:		SEASON:	○ SPRING ○ SUMMER ○ FALL ○ WINTER		
WEATHER CONDITIONS:		○ HOT ○ WARM ○ SUNNY ○ CLOUDY ○ RAINY ○ WINDY ○ FOGGY ○ COLD					
BUG NAME:							
SCIENTIFIC NAME:							
WHERE DID YOU FIND IT?							
WHAT COLOR(S) IS THE BUG?							
NUMBER OF LEGS?			DOES IT HAVE WINGS?		○ YES ○ NO ○ NOT SURE		
THE BUG IS...		○ BIG ○ SHINY ○ FAST ○ SCARY ○ LITTLE ○ SLOW ○ CUTE ○ ROUND ○ THIN					
DOES IT MAKE ANY SOUND?		○ YES ○ NO	WAS IT ALONE OR IN A GROUP?			○ ALONE ○ GROUP	

NOTES

PHOTO/DRAWING

BUG IDENTIFICATION

DATE:		TIME:		SEASON:	○ SPRING ○ SUMMER ○ FALL ○ WINTER

WEATHER CONDITIONS:	○ HOT ○ WARM ○ SUNNY ○ CLOUDY ○ RAINY ○ WINDY ○ FOGGY ○ COLD

BUG NAME:	
SCIENTIFIC NAME:	
WHERE DID YOU FIND IT?	
WHAT COLOR(S) IS THE BUG?	

NUMBER OF LEGS?		DOES IT HAVE WINGS?	○ YES ○ NO ○ NOT SURE

THE BUG IS...	○ BIG ○ SHINY ○ FAST ○ SCARY ○ LITTLE ○ SLOW ○ CUTE ○ ROUND ○ THIN

DOES IT MAKE ANY SOUND?	○ YES ○ NO	WAS IT ALONE OR IN A GROUP?	○ ALONE ○ GROUP

NOTES

PHOTO/DRAWING

BUG IDENTIFICATION

DATE:		TIME:		SEASON:	○ SPRING ○ SUMMER ○ FALL ○ WINTER		

WEATHER CONDITIONS:	○ HOT ○ WARM ○ SUNNY ○ CLOUDY ○ RAINY ○ WINDY ○ FOGGY ○ COLD

BUG NAME:	
SCIENTIFIC NAME:	
WHERE DID YOU FIND IT?	
WHAT COLOR(S) IS THE BUG?	

NUMBER OF LEGS?		DOES IT HAVE WINGS?	○ YES ○ NO ○ NOT SURE

THE BUG IS...	○ BIG ○ SHINY ○ FAST ○ SCARY ○ LITTLE ○ SLOW ○ CUTE ○ ROUND ○ THIN

DOES IT MAKE ANY SOUND?	○ YES ○ NO	WAS IT ALONE OR IN A GROUP?	○ ALONE ○ GROUP

NOTES

PHOTO / DRAWING

BUG IDENTIFICATION

DATE:		TIME:		SEASON:	○ SPRING ○ SUMMER ○ FALL ○ WINTER

WEATHER CONDITIONS:	○ HOT ○ WARM ○ SUNNY ○ CLOUDY ○ RAINY ○ WINDY ○ FOGGY ○ COLD

BUG NAME:	
SCIENTIFIC NAME:	
WHERE DID YOU FIND IT?	
WHAT COLOR(S) IS THE BUG?	

NUMBER OF LEGS?		DOES IT HAVE WINGS?		○ YES ○ NO ○ NOT SURE

THE BUG IS...	○ BIG ○ SHINY ○ FAST ○ SCARY ○ LITTLE ○ SLOW ○ CUTE ○ ROUND ○ THIN

DOES IT MAKE ANY SOUND?	○ YES ○ NO	WAS IT ALONE OR IN A GROUP?	○ ALONE ○ GROUP

NOTES

PHOTO / DRAWING

BUG IDENTIFICATION

DATE:		TIME:		SEASON:	○ SPRING ○ SUMMER ○ FALL ○ WINTER
WEATHER CONDITIONS:			○ HOT ○ WARM ○ SUNNY ○ CLOUDY ○ RAINY ○ WINDY ○ FOGGY ○ COLD		
BUG NAME:					
SCIENTIFIC NAME:					
WHERE DID YOU FIND IT?					
WHAT COLOR(S) IS THE BUG?					
NUMBER OF LEGS?		DOES IT HAVE WINGS?		○ YES ○ NO ○ NOT SURE	
THE BUG IS...	○ BIG ○ SHINY ○ FAST ○ SCARY ○ LITTLE ○ SLOW ○ CUTE ○ ROUND ○ THIN				
DOES IT MAKE ANY SOUND?	○ YES ○ NO	WAS IT ALONE OR IN A GROUP?		○ ALONE ○ GROUP	

NOTES

PHOTO / DRAWING

BUG IDENTIFICATION

DATE:		TIME:		SEASON:	○ SPRING ○ SUMMER ○ FALL ○ WINTER

WEATHER CONDITIONS:	○ HOT ○ WARM ○ SUNNY ○ CLOUDY ○ RAINY ○ WINDY ○ FOGGY ○ COLD

BUG NAME:	
SCIENTIFIC NAME:	
WHERE DID YOU FIND IT?	
WHAT COLOR(S) IS THE BUG?	

NUMBER OF LEGS?		DOES IT HAVE WINGS?	○ YES ○ NO ○ NOT SURE

THE BUG IS...	○ BIG ○ SHINY ○ FAST ○ SCARY ○ LITTLE ○ SLOW ○ CUTE ○ ROUND ○ THIN

DOES IT MAKE ANY SOUND?	○ YES ○ NO	WAS IT ALONE OR IN A GROUP?	○ ALONE ○ GROUP

NOTES

PHOTO / DRAWING

BUG IDENTIFICATION

DATE:		TIME:		SEASON:	○ SPRING ○ SUMMER ○ FALL ○ WINTER

WEATHER CONDITIONS:	○ HOT ○ WARM ○ SUNNY ○ CLOUDY ○ RAINY ○ WINDY ○ FOGGY ○ COLD
BUG NAME:	
SCIENTIFIC NAME:	
WHERE DID YOU FIND IT?	
WHAT COLOR(S) IS THE BUG?	

NUMBER OF LEGS?		DOES IT HAVE WINGS?	○ YES ○ NO ○ NOT SURE

THE BUG IS...	○ BIG ○ SHINY ○ FAST ○ SCARY ○ LITTLE ○ SLOW ○ CUTE ○ ROUND ○ THIN

DOES IT MAKE ANY SOUND?	○ YES ○ NO	WAS IT ALONE OR IN A GROUP?	○ ALONE ○ GROUP

NOTES

PHOTO / DRAWING

BUG IDENTIFICATION

DATE:		TIME:		SEASON:	○ SPRING ○ SUMMER ○ FALL ○ WINTER		
WEATHER CONDITIONS:		○ HOT ○ WARM ○ SUNNY ○ CLOUDY ○ RAINY ○ WINDY ○ FOGGY ○ COLD					
BUG NAME:							
SCIENTIFIC NAME:							
WHERE DID YOU FIND IT?							
WHAT COLOR(S) IS THE BUG?							
NUMBER OF LEGS?		**DOES IT HAVE WINGS?**		○ YES ○ NO ○ NOT SURE			
THE BUG IS...		○ BIG ○ SHINY ○ FAST ○ SCARY ○ LITTLE ○ SLOW ○ CUTE ○ ROUND ○ THIN					
DOES IT MAKE ANY SOUND?	○ YES ○ NO	**WAS IT ALONE OR IN A GROUP?**		○ ALONE ○ GROUP			

NOTES

PHOTO / DRAWING

BUG IDENTIFICATION

DATE:		TIME:		SEASON:	○ SPRING ○ SUMMER ○ FALL ○ WINTER		
WEATHER CONDITIONS:		○ HOT ○ WARM ○ SUNNY ○ CLOUDY ○ RAINY ○ WINDY ○ FOGGY ○ COLD					
BUG NAME:							
SCIENTIFIC NAME:							
WHERE DID YOU FIND IT?							
WHAT COLOR(S) IS THE BUG?							
NUMBER OF LEGS?			DOES IT HAVE WINGS?		○ YES ○ NO ○ NOT SURE		
THE BUG IS...		○ BIG ○ SHINY ○ FAST ○ SCARY ○ LITTLE ○ SLOW ○ CUTE ○ ROUND ○ THIN					
DOES IT MAKE ANY SOUND?	○ YES ○ NO		WAS IT ALONE OR IN A GROUP?			○ ALONE ○ GROUP	

NOTES

PHOTO / DRAWING

BUG IDENTIFICATION

DATE:		TIME:		SEASON:	○ SPRING ○ SUMMER ○ FALL ○ WINTER

WEATHER CONDITIONS:	○ HOT ○ WARM ○ SUNNY ○ CLOUDY ○ RAINY ○ WINDY ○ FOGGY ○ COLD
BUG NAME:	
SCIENTIFIC NAME:	
WHERE DID YOU FIND IT?	
WHAT COLOR(S) IS THE BUG?	

NUMBER OF LEGS?		DOES IT HAVE WINGS?	○ YES ○ NO ○ NOT SURE

THE BUG IS...	○ BIG ○ SHINY ○ FAST ○ SCARY ○ LITTLE ○ SLOW ○ CUTE ○ ROUND ○ THIN

DOES IT MAKE ANY SOUND?	○ YES ○ NO	WAS IT ALONE OR IN A GROUP?	○ ALONE ○ GROUP

NOTES

PHOTO / DRAWING

BUG IDENTIFICATION

DATE:		TIME:		SEASON:	○ SPRING ○ SUMMER ○ FALL ○ WINTER
WEATHER CONDITIONS:			○ HOT ○ WARM ○ SUNNY ○ CLOUDY ○ RAINY ○ WINDY ○ FOGGY ○ COLD		
BUG NAME:					
SCIENTIFIC NAME:					
WHERE DID YOU FIND IT?					
WHAT COLOR(S) IS THE BUG?					
NUMBER OF LEGS?		DOES IT HAVE WINGS?	○ YES ○ NO ○ NOT SURE		
THE BUG IS...		○ BIG ○ SHINY ○ FAST ○ SCARY ○ LITTLE ○ SLOW ○ CUTE ○ ROUND ○ THIN			
DOES IT MAKE ANY SOUND?	○ YES ○ NO	WAS IT ALONE OR IN A GROUP?	○ ALONE ○ GROUP		

NOTES

PHOTO/DRAWING

BUG IDENTIFICATION

DATE:		TIME:		SEASON:	○ SPRING ○ SUMMER ○ FALL ○ WINTER

WEATHER CONDITIONS:	○ HOT ○ WARM ○ SUNNY ○ CLOUDY ○ RAINY ○ WINDY ○ FOGGY ○ COLD
BUG NAME:	
SCIENTIFIC NAME:	
WHERE DID YOU FIND IT?	
WHAT COLOR(S) IS THE BUG?	

NUMBER OF LEGS?		DOES IT HAVE WINGS?		○ YES ○ NO ○ NOT SURE
THE BUG IS...	○ BIG ○ SHINY ○ FAST ○ SCARY ○ LITTLE ○ SLOW ○ CUTE ○ ROUND ○ THIN			
DOES IT MAKE ANY SOUND?	○ YES ○ NO	WAS IT ALONE OR IN A GROUP?		○ ALONE ○ GROUP

NOTES

PHOTO / DRAWING

BUG IDENTIFICATION

DATE:		TIME:		SEASON:	○ SPRING ○ SUMMER ○ FALL ○ WINTER	
WEATHER CONDITIONS:		○ HOT ○ WARM ○ SUNNY ○ CLOUDY ○ RAINY ○ WINDY ○ FOGGY ○ COLD				
BUG NAME:						
SCIENTIFIC NAME:						
WHERE DID YOU FIND IT?						
WHAT COLOR(S) IS THE BUG?						
NUMBER OF LEGS?			DOES IT HAVE WINGS?		○ YES ○ NO ○ NOT SURE	
THE BUG IS...		○ BIG ○ SHINY ○ FAST ○ SCARY ○ LITTLE ○ SLOW ○ CUTE ○ ROUND ○ THIN				
DOES IT MAKE ANY SOUND?	○ YES ○ NO		WAS IT ALONE OR IN A GROUP?		○ ALONE ○ GROUP	

NOTES

PHOTO/ DRAWING

BUG IDENTIFICATION

DATE:		TIME:		SEASON:	○ SPRING ○ SUMMER ○ FALL ○ WINTER

WEATHER CONDITIONS:	○ HOT ○ WARM ○ SUNNY ○ CLOUDY ○ RAINY ○ WINDY ○ FOGGY ○ COLD

BUG NAME:	
SCIENTIFIC NAME:	
WHERE DID YOU FIND IT?	
WHAT COLOR(S) IS THE BUG?	

NUMBER OF LEGS?		DOES IT HAVE WINGS?	○ YES ○ NO ○ NOT SURE

THE BUG IS...	○ BIG ○ SHINY ○ FAST ○ SCARY ○ LITTLE ○ SLOW ○ CUTE ○ ROUND ○ THIN

DOES IT MAKE ANY SOUND?	○ YES ○ NO	WAS IT ALONE OR IN A GROUP?	○ ALONE ○ GROUP

NOTES

PHOTO / DRAWING

BUG IDENTIFICATION

DATE:		TIME:		SEASON:	○ SPRING ○ SUMMER ○ FALL ○ WINTER		
WEATHER CONDITIONS:		○ HOT ○ WARM ○ SUNNY ○ CLOUDY ○ RAINY ○ WINDY ○ FOGGY ○ COLD					
BUG NAME:							
SCIENTIFIC NAME:							
WHERE DID YOU FIND IT?							
WHAT COLOR(S) IS THE BUG?							
NUMBER OF LEGS?			**DOES IT HAVE WINGS?**		○ YES ○ NO ○ NOT SURE		
THE BUG IS...		○ BIG ○ SHINY ○ FAST ○ SCARY ○ LITTLE ○ SLOW ○ CUTE ○ ROUND ○ THIN					
DOES IT MAKE ANY SOUND?	○ YES ○ NO		**WAS IT ALONE OR IN A GROUP?**		○ ALONE ○ GROUP		

NOTES

PHOTO / DRAWING

BUG IDENTIFICATION

DATE:		TIME:		SEASON:	○ SPRING ○ SUMMER ○ FALL ○ WINTER

WEATHER CONDITIONS:	○ HOT ○ WARM ○ SUNNY ○ CLOUDY ○ RAINY ○ WINDY ○ FOGGY ○ COLD

BUG NAME:	
SCIENTIFIC NAME:	
WHERE DID YOU FIND IT?	
WHAT COLOR(S) IS THE BUG?	

NUMBER OF LEGS?		DOES IT HAVE WINGS?	○ YES ○ NO ○ NOT SURE

THE BUG IS...	○ BIG ○ SHINY ○ FAST ○ SCARY ○ LITTLE ○ SLOW ○ CUTE ○ ROUND ○ THIN

DOES IT MAKE ANY SOUND?	○ YES ○ NO	WAS IT ALONE OR IN A GROUP?	○ ALONE ○ GROUP

NOTES

PHOTO / DRAWING

BUG IDENTIFICATION

DATE:		TIME:		SEASON:	○ SPRING ○ SUMMER ○ FALL ○ WINTER		
WEATHER CONDITIONS:		○ HOT ○ WARM ○ SUNNY ○ CLOUDY ○ RAINY ○ WINDY ○ FOGGY ○ COLD					
BUG NAME:							
SCIENTIFIC NAME:							
WHERE DID YOU FIND IT?							
WHAT COLOR(S) IS THE BUG?							
NUMBER OF LEGS?		**DOES IT HAVE WINGS?**		○ YES ○ NO ○ NOT SURE			
THE BUG IS...		○ BIG ○ SHINY ○ FAST ○ SCARY ○ LITTLE ○ SLOW ○ CUTE ○ ROUND ○ THIN					
DOES IT MAKE ANY SOUND?	○ YES ○ NO	**WAS IT ALONE OR IN A GROUP?**		○ ALONE ○ GROUP			

NOTES

PHOTO / DRAWING

BUG IDENTIFICATION

DATE:		TIME:		SEASON:	○ SPRING ○ SUMMER ○ FALL ○ WINTER

WEATHER CONDITIONS:	○ HOT ○ WARM ○ SUNNY ○ CLOUDY ○ RAINY ○ WINDY ○ FOGGY ○ COLD
BUG NAME:	
SCIENTIFIC NAME:	
WHERE DID YOU FIND IT?	
WHAT COLOR(S) IS THE BUG?	

NUMBER OF LEGS?		DOES IT HAVE WINGS?		○ YES ○ NO ○ NOT SURE
THE BUG IS...	○ BIG ○ SHINY ○ FAST ○ SCARY ○ LITTLE ○ SLOW ○ CUTE ○ ROUND ○ THIN			
DOES IT MAKE ANY SOUND?	○ YES ○ NO	WAS IT ALONE OR IN A GROUP?		○ ALONE ○ GROUP

NOTES

PHOTO / DRAWING

BUG IDENTIFICATION

DATE:		TIME:		SEASON:	○ SPRING ○ SUMMER ○ FALL ○ WINTER

WEATHER CONDITIONS:	○ HOT ○ WARM ○ SUNNY ○ CLOUDY ○ RAINY ○ WINDY ○ FOGGY ○ COLD

BUG NAME:	
SCIENTIFIC NAME:	
WHERE DID YOU FIND IT?	
WHAT COLOR(S) IS THE BUG?	

NUMBER OF LEGS?		DOES IT HAVE WINGS?	○ YES ○ NO ○ NOT SURE

THE BUG IS...	○ BIG ○ SHINY ○ FAST ○ SCARY ○ LITTLE ○ SLOW ○ CUTE ○ ROUND ○ THIN

DOES IT MAKE ANY SOUND?	○ YES ○ NO	WAS IT ALONE OR IN A GROUP?	○ ALONE ○ GROUP

NOTES

PHOTO/DRAWING

BUG IDENTIFICATION

DATE:		TIME:		SEASON:	○ SPRING ○ SUMMER ○ FALL ○ WINTER

WEATHER CONDITIONS:	○ HOT ○ WARM ○ SUNNY ○ CLOUDY ○ RAINY ○ WINDY ○ FOGGY ○ COLD

BUG NAME:	
SCIENTIFIC NAME:	
WHERE DID YOU FIND IT?	
WHAT COLOR(S) IS THE BUG?	

NUMBER OF LEGS?		DOES IT HAVE WINGS?	○ YES ○ NO ○ NOT SURE

THE BUG IS...	○ BIG ○ SHINY ○ FAST ○ SCARY ○ LITTLE ○ SLOW ○ CUTE ○ ROUND ○ THIN

DOES IT MAKE ANY SOUND?	○ YES ○ NO	WAS IT ALONE OR IN A GROUP?	○ ALONE ○ GROUP

NOTES

PHOTO/DRAWING

BUG IDENTIFICATION

DATE:		TIME:		SEASON:	○ SPRING ○ SUMMER ○ FALL ○ WINTER

WEATHER CONDITIONS:	○ HOT ○ WARM ○ SUNNY ○ CLOUDY ○ RAINY ○ WINDY ○ FOGGY ○ COLD

BUG NAME:	
SCIENTIFIC NAME:	
WHERE DID YOU FIND IT?	
WHAT COLOR(S) IS THE BUG?	

NUMBER OF LEGS?		DOES IT HAVE WINGS?	○ YES ○ NO ○ NOT SURE

THE BUG IS...	○ BIG ○ SHINY ○ FAST ○ SCARY ○ LITTLE ○ SLOW ○ CUTE ○ ROUND ○ THIN

DOES IT MAKE ANY SOUND?	○ YES ○ NO	WAS IT ALONE OR IN A GROUP?	○ ALONE ○ GROUP

NOTES

PHOTO / DRAWING

BUG IDENTIFICATION

DATE:		TIME:		SEASON:	○ SPRING ○ SUMMER ○ FALL ○ WINTER

WEATHER CONDITIONS:	○ HOT ○ WARM ○ SUNNY ○ CLOUDY ○ RAINY ○ WINDY ○ FOGGY ○ COLD

BUG NAME:	
SCIENTIFIC NAME:	
WHERE DID YOU FIND IT?	
WHAT COLOR(S) IS THE BUG?	

NUMBER OF LEGS?		DOES IT HAVE WINGS?	○ YES ○ NO ○ NOT SURE

THE BUG IS...	○ BIG ○ SHINY ○ FAST ○ SCARY ○ LITTLE ○ SLOW ○ CUTE ○ ROUND ○ THIN

DOES IT MAKE ANY SOUND?	○ YES ○ NO	WAS IT ALONE OR IN A GROUP?	○ ALONE ○ GROUP

NOTES

PHOTO/DRAWING

BUG IDENTIFICATION

DATE:		TIME:		SEASON:	○ SPRING ○ SUMMER ○ FALL ○ WINTER		
WEATHER CONDITIONS:			○ HOT ○ WARM ○ SUNNY ○ CLOUDY ○ RAINY ○ WINDY ○ FOGGY ○ COLD				
BUG NAME:							
SCIENTIFIC NAME:							
WHERE DID YOU FIND IT?							
WHAT COLOR(S) IS THE BUG?							
NUMBER OF LEGS?			DOES IT HAVE WINGS?		○ YES ○ NO ○ NOT SURE		
THE BUG IS...			○ BIG ○ SHINY ○ FAST ○ SCARY ○ LITTLE ○ SLOW ○ CUTE ○ ROUND ○ THIN				
DOES IT MAKE ANY SOUND?		○ YES ○ NO	WAS IT ALONE OR IN A GROUP?		○ ALONE ○ GROUP		

NOTES

PHOTO / DRAWING

BUG IDENTIFICATION

DATE:		TIME:		SEASON:	○ SPRING ○ SUMMER ○ FALL ○ WINTER		
WEATHER CONDITIONS:		○ HOT ○ WARM ○ SUNNY ○ CLOUDY ○ RAINY ○ WINDY ○ FOGGY ○ COLD					
BUG NAME:							
SCIENTIFIC NAME:							
WHERE DID YOU FIND IT?							
WHAT COLOR(S) IS THE BUG?							
NUMBER OF LEGS?			DOES IT HAVE WINGS?		○ YES ○ NO ○ NOT SURE		
THE BUG IS...		○ BIG ○ SHINY ○ FAST ○ SCARY ○ LITTLE ○ SLOW ○ CUTE ○ ROUND ○ THIN					
DOES IT MAKE ANY SOUND?		○ YES ○ NO	WAS IT ALONE OR IN A GROUP?		○ ALONE ○ GROUP		

NOTES

PHOTO / DRAWING

CPSIA information can be obtained
at www.ICGtesting.com
Printed in the USA
LVHW011524191220
674611LV00004B/1011